PREFACE

In the 1960s, Dr Isles Strachan, then of the University of E
visitors' book which had been uncovered in an outhouse at the ....
looking through it he found the signatures of groups of famous geologists who had visited
Assynt early this century. He took a photograph of one of the pages, passing it later to his
colleague Dr Roy MacGregor.

In April 1993, when Dr MacGregor and Sinclair Ross were 'preparing the ground' for the
Edinburgh Geological Society's long excursion to Assynt, they enquired at the hotel
about the book and found that it was still in existence, though in poor condition. They
arranged for it to be put on display when the main party came the following month.

The party viewed it with such interest that an approach was subsequently made to the
hotel's proprietors to have the book refurbished at the Society's expense.

With the book neatly restored, a small group of Fellows of the Society then took it back
to the hotel, where the President presented it to the proprietors along with a
commemorative plaque and two framed photographs. The plaque outlines the historical
importance of Assynt to the international geological community, and of the Inchnadamph
Hotel as a centre for visiting geologists. One of the photographs shows two pages from
the hotel register bearing the names of the thirty geologists who attended the British
Association's excursion to Assynt in September 1912, led by Ben Peach and John Horne
of the British Geological Survey. Those present included many of the foremost
international authorities on mountain-building processes, and three future Directors of the
British Geological Survey. The other photograph is an aerial view of part of the Moine
Trust Zone near the hotel.

The plaque and photographs can be seen in the hotel foyer.

Today, geologists still come from all over the world to admire and study the geology of
Assynt, and this booklet is a tribute to those brilliant and dedicated pioneers who first
unravelled its complexities. Fishermen, walkers and tourists are also drawn to Assynt,
and I hope that these pages will add to their appreciation of this wild and beautiful area.

S Ian Hogarth
President
Edinburgh Geological Society

October 1995

ASSYNT

The geologists' Mecca

Researched and produced on behalf of the
Edinburgh Geological Society

by

P M Dryburgh
A R MacGregor
S M Ross
C L Thompson

CONTENTS

PLATES

THE GEOLOGY OF ASSYNT

The scenery

The Inchnadamph Hotel, Assynt, lies in the centre of a landscape which, in its austerity and grandeur, is unique in the British Isles. The lower ground around Lochinver is hummocky with stretches of peat bog and water lying between bare rocky knolls. Rising from this foundation are isolated mountains, for example Canisp and Suilven. Some have summits like castle walls; others are capped by a white rock which gleams almost like snow in the sunlight.

The rocks which produce such magnificent scenery occupy a strip of country about 15-25km wide and over 150km long. It extends from the north coast around Cape Wrath and Loch Eriboll southwards past Assynt, Ullapool and Kyle of Lochalsh to the Sleat Peninsula in Skye. Within this strip it is the Assynt District which contains some of the finest scenery and most spectacular rock formations.

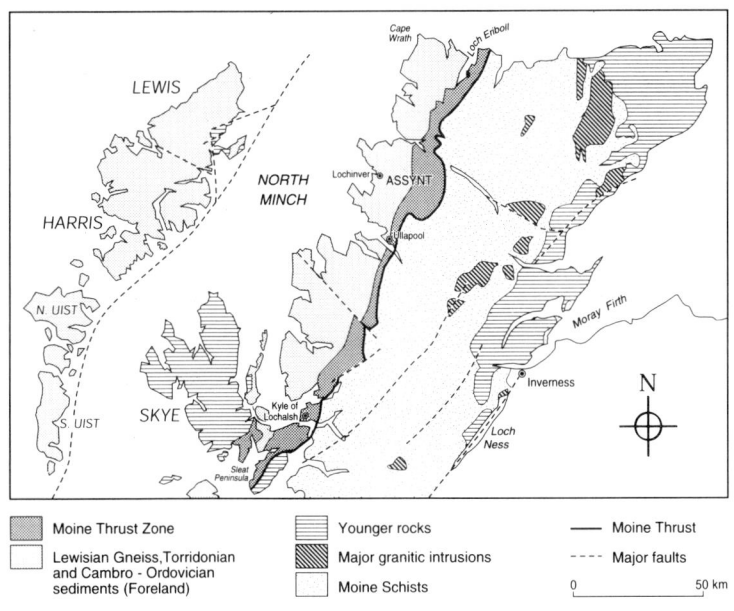

Figure 1: Sketch map showing the main elements of the geology of the Northern Highlands.

The rocks beneath

To see how this landscape was formed, we have to make use of the science of geology, the study of the structure, composition and history of the Earth. Geology is chiefly concerned with the Earth's outermost shell, or crust, where the rock layers, or beds, are usually piled on top of each other in the order in which they were formed. As a result the oldest layers are at the bottom and the youngest at the top, *unless they have been disturbed.* This simple but important rule is known as the Law of Superposition. By examining the many exposures of bare rock in the Northwest Highlands and using the Law, geologists have worked out, for that area, the general sequence of rocks from bottom to top, and their relative ages. (Figure 2.)

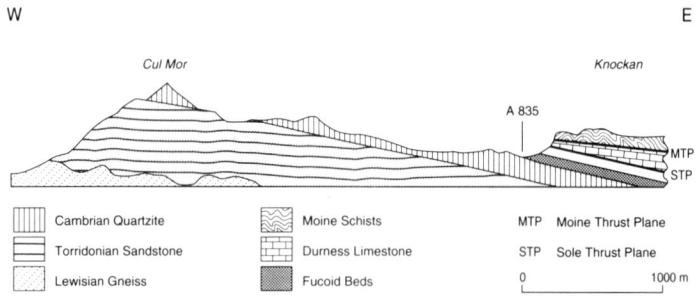

Figure 2: Cross-section through Cul Mor and Knockan Cliff showing older Moine Schists thrust over younger Cambro-Ordovician rocks. (MTP = Moine Thrust Plane: STP = Sole Thrust Plane

At the base of the sequence, in the west, are the rocks giving the low, hummocky terrain. These rocks are complex and very old - the oldest in the British Isles. Early in the Earth's history they were altered by being buried to great depth, heated and recrystallized to form a kind of metamorphic (changed) rock called *gneiss*. It is recognisable by its crystalline appearance with grey and pink bands. The same rock makes up so much of the Isle of Lewis in the Outer Hebrides that it is called the Lewisian Gneiss, or Lewisian for short.

Resting on the Lewisian is the dark, red-brown, Torridonian Sandstone, named after the Torridon district of Wester Ross, where it forms the spectacular Torridon mountains. It occurs in near-horizontal beds up to 2m thick. These beds, cut by very noticeable vertical joints, give the mountains their characteristic ledges and steep cliffs.

At one time the Torridonian buried the Lewisian to a depth of about 7km. As hundreds of millions of years went by, however, it was uplifted and then worn away by the combined effects of weathering and erosion. Now all that is left of it in the Assynt area are the isolated masses of Canisp, Suilven, Cul Mor, Cul Beag, Stac Pollaidh and Quinag.

In contrast with the Lewisian, the Torridonian is undisturbed and unmetamorphosed. It rests directly on the Lewisian and is therefore the younger of the two.

Lying on the Torridonian, and younger still, are beds of a white sandstone, hardened into a rock called quartzite. The quartzite forms the snow-like summits of mountains such as Canisp and Quinag. Above the quartzite is a mixed sequence of rocks including some once-muddy layers called the Fucoid Beds. These beds were of great importance in the later history of the rocks, as will be seen.

The succession then continues upwards with beds of the Durness Limestone. It is this limestone that is responsible for the caves, swallow-holes and underground streams east of Inchnadamph and elsewhere. It also supports an unusual community of lime-loving plants.

Stretching 50 to 100km to the east of all these rocks are the Moine Schists, which take their name from the A'Mhoine (*Gaelic - peat or bog*) peninsula in northern Sutherland. Schist is a metamorphic rock that splits along bands of tiny flakes of the mineral mica. The Moine Schists give rise to the rather bleak moorland landscape to the east of the Northwest Highlands.

In fact, each of the geological units listed above produces its own characteristic scenery, with the result that an experienced geologist can, sitting on a hilltop, read the geology of the country using binoculars. This approach was first demonstrated in the Swiss Alps and led to geologists referring to binoculars as '*Schweitzerhammer*'. In Assynt, individual formations such as the white Cambrian quartzite can sometimes be followed for miles across country or from peak to peak.

A geological puzzle

From their position on the top of the sequence of rocks just described, the Moine Schists appear to be the youngest rocks in the Northwest Highlands. Yet, like the Lewisian, they too are metamorphosed, implying that they have at one time been buried deep inside the Earth's crust. It is hard to imagine how they could have suffered this treatment while the apparently older rocks sandwiched between them and the Lewisian Gneiss have not.

For most of last century a succession of geologists battled over the answer to this geological puzzle. Were the Moine Schists really younger than the other rocks in the area? Or were they another form of the Lewisian Gneiss? In which case, why were they above rocks which themselves lay on top of the Lewisian?

The answer, when it came in the 1880s, was a breakthrough in the understanding of Highland geology. The Moine Schists could not, for certain, be said to be younger than the Cambro-Ordovician because, contrary to the key phrase in the Law of Superposition, *they had been disturbed*. They had been transported from the place in which they had been formed to what is now the Northwest Highlands.

Just how great that movement had been, and how it had taken place, was the revelation that shook the geological world. For it turned out that the Moine Schists had formed far away, many kilometres to the east. They had been driven - or thrust - westwards, while several kilometres deep inside the crust, on gently inclined surfaces called thrust planes. They had finally come to rest on top of the Cambro-Ordovician and older rocks of the Northwest Highlands, so giving the rock sequence which puzzled so many geologists for so long (Figure 3).

The most important of the thrusts is the Moine Thrust which can be traced from Loch Eriboll to Skye. Several other major thrust planes were also found, together with many minor ones. In effect these thrusts had cut the Earth's crust into many slices, large and small, making the pattern of rocks at ground level very complicated. This "Zone of Complication", as the early geologists called it, is now referred to as the Moine Thrust Zone. Its position is shown in Figure 1.

As for the age of the Moine Schists, that remained in dispute until recently when it was established that they were older than the Torridonian (Figure 3).

The sequence of rocks described so far can be summarised in a diagram with the oldest rocks at the bottom and the youngest at the top.

Figure 3 is such a diagram showing the sequence of rocks which occurs in Assynt and it indicates the names (Precambrian, Cambrian and Ordovician) of the periods of Earth history when the rocks were formed. On the left of Figure 3 are the corresponding ages in millions of years (Ma).

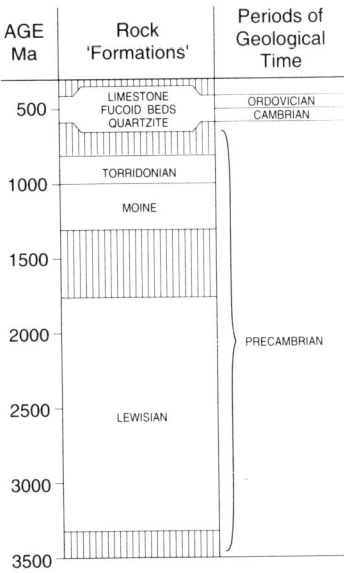

Figure 3: Summary of the rock sequence in Assynt

HOW THRUSTING WORKS

Thrust faults

Thrusts on the scale seen in the Northwest Highlands are formed when mountain chains are being built. In the infant mountain belt, immense, slow earth movements begin, exerting great pressure on the rocks and squeezing them as if they were between the jaws of a vice. The rocks react to this pressure, which is largely horizontal, by both buckling and fracturing. The affected area shortens in the direction in which the pressure is applied and at the same time thickens, so that the mountain chain not only rises but also develops deep roots.

To see how this happens, imagine some beds of rock lying flat and undisturbed between the jaws of a large vice, as in Figure 4a, below. Bed A is older than B which is older than C.

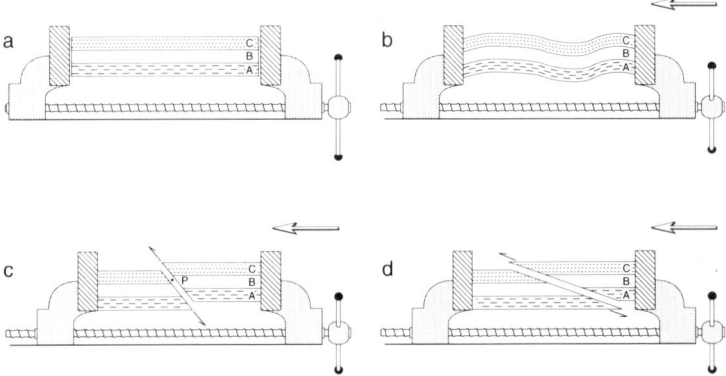

Figure 4: Deformation under horizontal pressure. (After Cadell, 1889).

If the rocks are compressed by moving the right-hand jaw towards the left, the rocks may fold, as in Figure 4b.

Alternatively they may fracture into two or more blocks, with one block riding up over another as in Figure 4c. This kind of structure is called a *fault*, and the plane of fracture, a *fault plane*. Notice how older rocks come to rest on younger, as at point P, where bed B rests on bed C.

If the fault plane is at a low angle to the horizontal, as in Figure 4d, the structure is called a *thrust fault*.

Thrust faults played a key role in the formation of the Northwest Highlands and so they deserve further explanation.

Piggyback thrusting

When a major thrust plane forms in a developing mountain belt, the fracture cuts upwards through the many layers of rock in a gentle curve as in Figure 5a. The rocks on the upper side of the fault plane then move as a single large sheet, riding up the fault plane and over the stationary rocks beneath it. The moving rocks can travel many kilometres from where they started, and old rocks from deep down can finish up lying on much younger rocks (Figure 5b).

Eventually an obstruction may bring the sheet, Sheet 1, to a halt, but if pressure continues, a new fault, Thrust 2, may form below and ahead of the first. The rocks above Thrust 2 then move as a second sheet, Sheet 2, carrying Sheet 1 and Thrust 1 as passive passengers, piggyback-style.

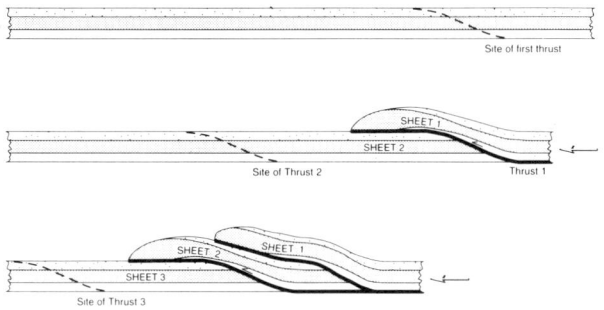

Figure 5: Piggyback thrusting.

Later on, Sheet 2 may also come to a halt, and yet another thrust, Thrust 3, may form beneath it, carrying sheets 1 and 2, and their thrusts piggyback. By this process of "piggyback thrusting" the Earth's crust is shortened and thickened in the zone of thrusting as multilayered sheets are piled one on top of the other.

A closer look at the side view of a major thrust will reveal that the line of fracture is often not a smooth curve but stepped, like a staircase. The treads of the staircase are called *flats* and occur in weak layers of rock, such as clays or shales. The steeper parts are called *ramps* and occur where the thrust cuts up through stronger layers such as sandstones.

In the manner described above, but on a much smaller scale, a succession of small slices are thrust up as the ramping progresses. These do not override, and as a result, the layer is repeatedly sliced and stacked up like a pile of roofing-tiles. This stacking, termed *imbrication*, usually occurs under an overlying thrust sheet and pushes it up into a dome (Figure 6).

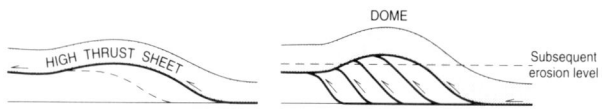

Figure 6: Imbrication stacking causing doming. Erosion of dome results in a 'window' which exposes the structures below.

It is only in the last three decades, as a result of the exploration for oil, that the "rules" which thrusting follows have been worked out. Yet as early as 1889 the Scottish geologist, H M Cadell, was able to demonstrate how thrusting works (Plate 1) by compressing layers of wet sand and plaster of Paris in a pressure box,

The Mechanics of Thrusting

It is not easy to visualise one sheet of rock, weighing perhaps billions of tonnes, "gliding" over another. After all it is difficult enough for a human to push a paving slab along the ground. One has to remember, though, that thrusting often takes place many kilometres down in the Earth's crust. There, the water trapped in the pores of the rocks is at such high pressure that it is able to carry the load of the overlying rocks. They can then move with relative ease, supported on a cushion of water. Even so, the rocks at the base of the moving sheet can suffer intense deformation. At depths below about 15 km, this results in their being reconstituted as a hard, streaky rock called *mylonite*.

Thrusting in Assynt

Let us turn now to the Northwest Highlands, where we can see the characteristics of thrust-fault zones - piggyback thrusting, imbrication, mylonites and so on - wonderfully displayed. The mountains of the Northwest Highlands form the boundary of a mountain chain which began to form about 500 million years ago. To the west of the Moine Thrust Zone, the Lewisian, Torridonian and Cambro-Ordovician rocks behaved like the stationary jaw of the vice, forming a stable block known as a *foreland*. Far away to the east, earth movements began, pushing westwards like the moving jaw of the vice. This folded the Moine Schists and then thrust them westwards as a sheet along the Moine Thrust Plane and onto the foreland. As compression continued, other major thrusts developed deeper down, taking the easy route provided by weak rocks, such as the Fucoid Beds. The rocks of the underlying foreland were thus sliced into sheets and, by the process of piggyback thrusting, piled up in a different order from that in which they were originally laid down.

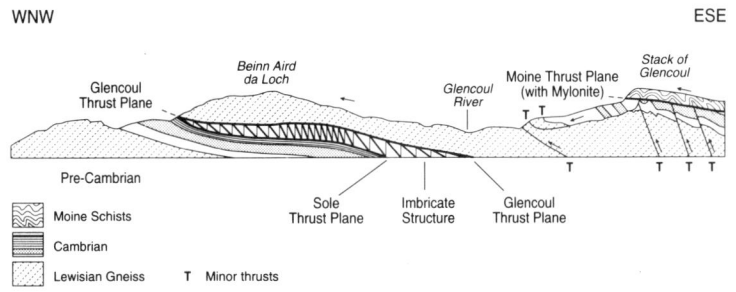

Figure 7: A vertical cross-section through Assynt.

Figure 7 shows the Moine Thrust Plane, the oldest of the major thrusts in Assynt, which brought Moine Schists on to the top of Cambro-Ordovician rocks. Its presence is marked in many places by mylonite, which shows that the thrust plane was originally 15 km or more deep. Below the Moine Thrust is the younger Glencoul Thrust which brought Lewisian on to the top of Cambro-Ordovician. Below that is the Sole Thrust, the youngest of the major Assynt thrusts, which carried the other, older thrusts piggyback. In the sheet that lies between the Glencoul and Sole Thrusts can be seen a whole succession of mini-thrusts which, one after the other, sliced the Cambro-Ordovician and stacked it into the imbricate or roofing-tile arrangement described earlier.

By carefully surveying and drawing sections of the Moine Thrust Zone, geologists have been able to "run the film backwards" and calculate how much slippage took place across the whole Zone. It was at least 77 km, and could have been over 100 km.

In Assynt there was such a pile-up of thrust slices, large and small, including some of the Lewisian, that the rocks above were domed upwards. During the hundreds of millions of years that followed, the area was uplifted and eroded. Eventually the top of the Assynt "dome" was stripped off, opening up a "window" on to the structures beneath. This is why Assynt is so important in the unravelling of the geological story of the Northwest Highlands.

Plate 3. Cadell's "squeeze-box" experiment (1889)

THE HIGHLANDS CONTROVERSY

Mountain-building — early ideas

Today's geologists fully appreciate the enormous length of time it takes to build a mountain chain, and are well aware that large horizontal as well as vertical movements are involved. Early geologists, however, were taught that mountain-building movements were mostly up and down. Consequently they did not recognise large thrusts and so did not understand the structure of the Northwest Highlands.

The interpretation of mountain structures in terms of thrust sheets (*nappes*) began in 1841 when the Swiss geologist, Albert Escher von der Linth, announced his discovery of older rocks lying over younger ones in the Alps. This he attributed to folding on a grand scale. The distance the rocks had travelled was so great that he was afraid to publish the discovery in case he was thought mad. However, after his death, Albert Heim, his successor as Professor of Geology at Zürich, did so in a classic textbook on mountain-building. At the same time other studies were being carried out in the USA and in Scandinavia.

The Highlands Controversy

The Geological Survey of Great Britain was formed in 1835 and was the first national geological survey of any country in the world. In 1860 a controversy over the structure of the Northwest Highlands developed between the 'professional' geologists of the young Geological Survey under their Director, Sir Roderick Murchison, and academic geologists, who were looked on as amateurs.

Murchison was an ex-soldier whose wife had persuaded him to take up geology as an outdoor substitute for fox-hunting. He had earlier recognised a new suite of old rocks in Wales, which he termed the Silurian. In keeping with his military past Murchison then tried to "conquer" large parts of the world for British science by labelling their rocks as Silurian too. He did this for the Southern Uplands of Scotland, but when he tried to add the Northwest Highlands to his Silurian empire, he was challenged by the 'amateur' Professor James Nicol of Aberdeen. Murchison thought the rocks in the Northwest became progressively younger upwards, while Nicol maintained that older rocks lay on top of younger ones, having been carried there on fault planes. A bitter dispute followed in which Murchison, with many friends in high places in the scientific community, saw to it that Nicol was discredited.

Murchison appointed his disciple, Archibald Geikie, as Director of the Geological Survey in Scotland, and together they formulated a general theory of the structure of Scotland based on Murchison's ideas. A few years later, the ambitious Geikie became, in addition, the first professor of geology at Edinburgh - a post set up by Murchison just before his death. Geikie's brilliance as a communicator ensured that Murchison's theories reached a very wide audience and were not challenged for many years (see plate 4).

Another amateur emerged, in the shape of Charles Lapworth, a schoolmaster with no formal education in geology. He commenced a geological study of a small area in the Southern Uplands just at the time the official survey of that part was starting under Geikie's direction. The 'professional' surveyors used the traditional methods of working out the geology by making superficial examinations of the rocks during rapid traverses of the countryside. Their progress became increasingly bogged down as they failed to recognise that the rock successions had been repeated many times by faulting and therefore appeared much thicker. Lapworth, on the other hand, carried out detailed large-scale mapping of small areas and successfully demonstrated that the same beds had been repeated many times. In the following ten years he described the sequences across the whole area, using the same methods. His work was received with wide acclaim and he was appointed professor of geology at Birmingham in 1881 as a result.

This was extremely embarrassing to Geikie, as the publication of maps and descriptions of the Survey's work to date had to be scrapped, and almost the whole of the area had to be resurveyed using Lapworth's methods.

The debate on the structure of the Northwest Highlands was then reopened by a succession of 'amateurs' who found evidence which supported Nicol's views. When, in 1880, Charles Calloway published evidence exposing errors in the official theory, Geikie himself journeyed north to review the situation.

It was Lapworth, however, who finally provided the solution, by recognising similarities between the problems in the Northwest and those in the Alps, the USA and Scandinavia. He started work in Eriboll, and soon realised that horizontal pressure had driven older rocks from the east for many miles over the top of younger ones to the west, pulverising and flattening some of them into laminated sheets. It was these laminations which Murchison had identified as bedding that made the succession appear to be continuous. This first recognition of thrust planes proved to be the key to the problem.

Lapworth published his first results in 1883, but broke down under the excitement of his discoveries, imaging the Great Moine Nappe grating over his body as he lay tossing in his bed. This delayed the appearance of his full report. Geikie, who was now the Director General of the Survey, sent his best surveyors, Ben Peach and John Horne, to the Northwest. The following year, they demonstrated to Geikie in the field that the 'amateurs' had again been correct. Geikie returned to London, and a few days later a brief report on the findings of Peach and Horne was published in *Nature*, with a preface by Geikie in which he announced that he now accepted the new interpretation. However, he made no acknowledgement at all that others such as Calloway and Lapworth had worked out the structures independently, and earlier than the Survey. When the official Survey report was published in 1888, Horne did acknowledge the work of the 'amateurs'. However Geikie did not, and this triggered a wave of resentment which culminated in an official enquiry into the efficiency of the Survey under his leadership, and to his eventual retirement.

The Geological Survey's programme of mapping of the Northwest Highlands had begun on the North coast in 1883 and worked southwards to Skye. The task was finally completed in 1897. While some five or six members of the Survey were responsible for most of the mapping, the major part of the work was carried out by Ben Peach and John Horne, and the final memoir is always associated with these two. This memoir, *"The Geological Structure of the Northwest Highlands of Scotland"*, appeared in 1907. It was largely written by John Horne and edited by Geikie. The structures along the entire length of the thrust zone are described in great detail in clear prose accompanied by excellent illustrations. To this day it remains a classic of geological literature, and it is no exaggeration to say that it is one of the most notable geological memoirs ever published in the English language. It makes a fitting memorial to the team who successfully tackled such a complicated survey - once they had been prompted by the 'amateurs'.

Progress elsewhere

Meanwhile, on the Continent, geologists were working on similar problems, and in 1882 in Stockholm, Brogger quoted some of Lapworth's methods in a textbook on the geology of the Oslo area. In 1884, in France, Marcel Bertrand suggested that the fold in the Alps, as described by Escher, was in fact a thrust. In 1888 Törnebohm showed that the Scandinavian mountain chain had overridden the foreland which lay on its southeast side by at least 100 km. In 1893 Schardt interpreted the northwestern foothills of the Alps near Lake Geneva as being a pile of far-travelled masses or nappes, now separated from their roots by a great distance.

The last act of the great drama came when Maurice Lugeon published his memoir on the structure of the Swiss Alps in 1902. His explanations of the structures in terms of nappes and thrust movements were masterly, and his

Deux géologues rapportant à Mrs Ellice, un petit échantillon des Montagnes moutonnées découvertes en 1860, entre. Glenquoich et Lochhourne

[en gaélique : Slabbgdh]

Plate 4. Cartoon of Murchison and Geikie by Prosper Mérimée (1860). Courtesy of the Trustees of the National Library of Scotland.

old professor, Albert Heim, expressed joy at the revelations contained in the work. As a postscript, Emil Haug, in 1903, discovered the missing thrusts needed to confirm Bertrand's theory.

Assynt - the geologists' Mecca

From the 1890s onwards there was a steady flow of geological visitors to the Northwest Highlands, particularly to Inchnadamph. Some of the most distinguished visitors came from abroad, including most of those mentioned above. Peach and Horne were much in demand to act as guides, a task which they willingly undertook for as long as they were able. Even today there is no let-up: geologists still travel to the Northwest Highlands from all over the world to marvel at the work of these pioneers, and to ponder.

THE BRITISH ASSOCIATION EXCURSION
TO ASSYNT, 1912

The 1907 Northwest Highland Memoir by Peach and Horne brought before a world-wide readership a lucid account of what was one of the most exciting areas of geological research. With the phenomena so well exposed on the ground, what could have been more natural, when the British Association for the Advancement of Science met in Dundee in September 1912, than for geologists from many parts of the world to take the opportunity of going to visit Assynt? After the meeting, thirty of the delegates embarked on a week-long excursion to Assynt, conducted by Ben Peach and John Horne.

They travelled by rail from Dundee via Perth and Inverness to Lairg, on the outward journey, and from Lairg to Edinburgh on the return. The party had a carriage specially reserved for the journey at a reduced rate for the round trip of the single fare plus a third, which came out at twenty shillings and sixpence (£1.03) per person. On the way north they broke their journey, staying overnight on Wednesday 11th at the Station Hotel in Inverness, where they were charged eleven shillings and sixpence (57p) for dinner, bedroom, breakfast and attendance (exclusive of spirits, wines and aerated waters). Those sharing a double room had a reduction of one shilling (5p). They arrived in Lairg the next day in time for lunch and then travelled on to Assynt by "auto-car". At the Inchnadamph Hotel the party was accommodated at the reduced rate of nine shillings (45p) per day for full board (exclusive of spirits, wines and aerated waters), with transport charges extra.

During their stay the party examined the Lewisian, Torridonian and Cambrian rocks of the foreland round Loch Assynt and Lochinver; and various sections within the Moine Thrust Zone near the hotel, on Beinn an Fhuarain, at the Knockan Cliff and in the Loch Ailsh/River Oykell area. They also made a boat trip from Kylesku to the head of Loch Glencoul to examine the structures at the Glencoul Thrust Plane, and younger members of the party climbed to the base of the Stack of Glencoul to see the mylonites associated with the Moine Thrust. The igneous rocks at Borrolan and Loch Ailsh and their local metamorphic effects were also included, but the main focus of the excursion was the Moine Thrust Zone. In all, it was an itinerary little different from that pursued today.

These arrangements must have involved a considerable amount of work, especially in view of the international nature of the party. Few other details of the excursion are known, but fortunately several photographs of the group were taken by Professor Reynolds. Two of these are reproduced in this booklet (inside front cover and plate 7). One delightful story that has survived was recounted by one of the members on his 100th birthday in 1989. Tresillian Nicholas, who had been a young Cambridge research assistant at the time, recalled that the party sat round the fire, in those pre-television days, singing in French "The Moine Thrust Song" on the final evening at the hotel. This song had been composed and set to music by another party member, Professor Lugeon.

Proceedings were brought to a close by Albert Heim, the father-figure present, who proposed a very moving vote of thanks to Peach and Horne for their leadership during the excursion:

"Ladies and Gentlemen,

We are now at the end of our beautiful excursion in the Highlands of Scotland.

Our predominant feeling and impulse is to thank our guides *Peach* and *Horne*!

We look at the scientific work they have done in this country with the highest respect and appreciation. They are a couple of scientists, Investigator-Twins, such as I never have seen before in my life, two men so delightfully developed in a wonderful common work of research."

Heim went on to describe how Peach and Horne had been working on the weathered-down roots of an ancient mountain chain and invited them to visit the Alps to see "the much younger leaves and beautifully folded flowers" which followed higher up in the sequence.

"... Come! You must come! It is your duty, it is your human and your scientific right. It will be a reward for your excellent life-work. Let me thank you in the name of science for your scientific work. It was a difficult work, a hard work, and a work which could only be done by men of your strength and devotion!"

"... We all will hold our Investigator Twins dear in our memory. May they live long and happy and enjoy the growth of knowledge and the respect and love of everybody. So let us thank you once more!"

The group's itinerary, the photographs by Professor Reynolds, a copy of "La Chanson du Moine Thrust" and Albert Heim's vote of thanks are preserved in the British Geological Survey's archives.

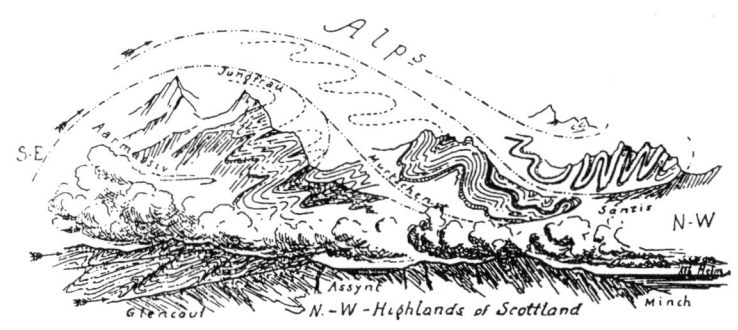

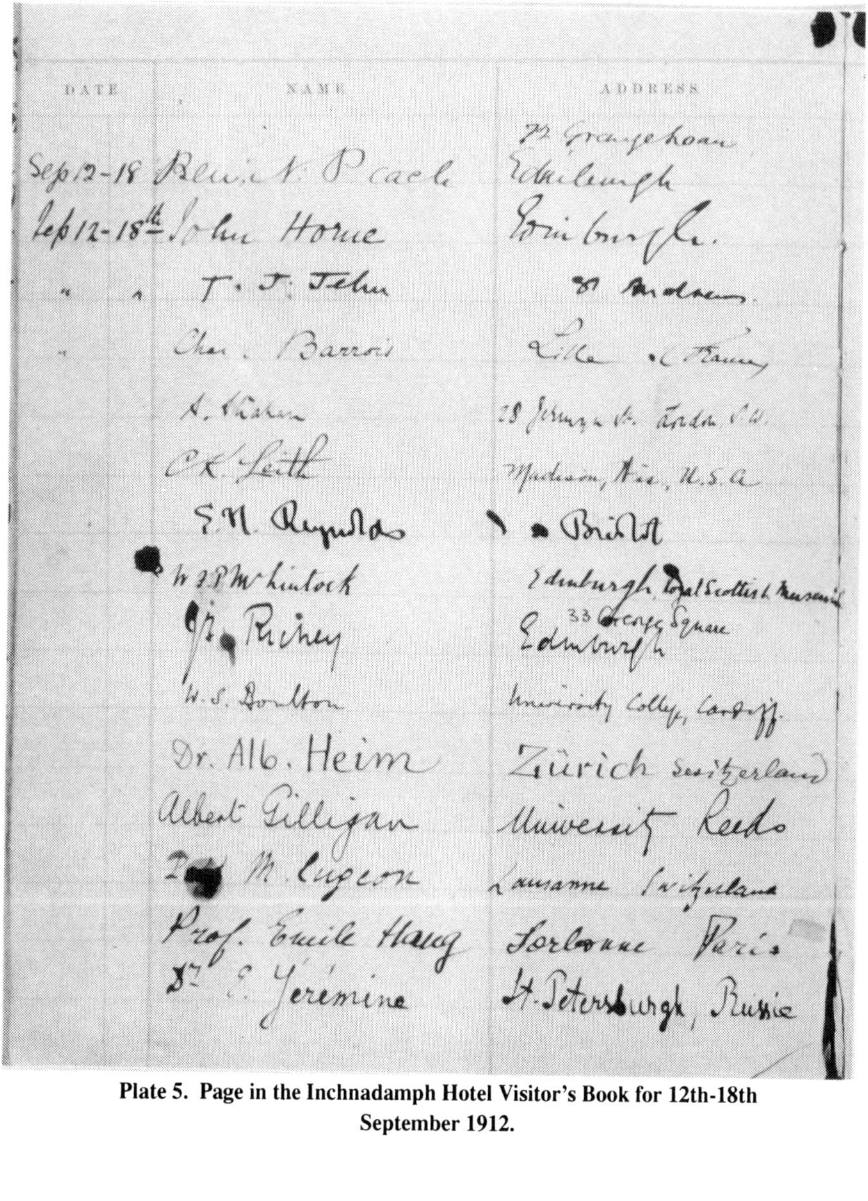

Plate 5. Page in the Inchnadamph Hotel Visitor's Book for 12th-18th September 1912.

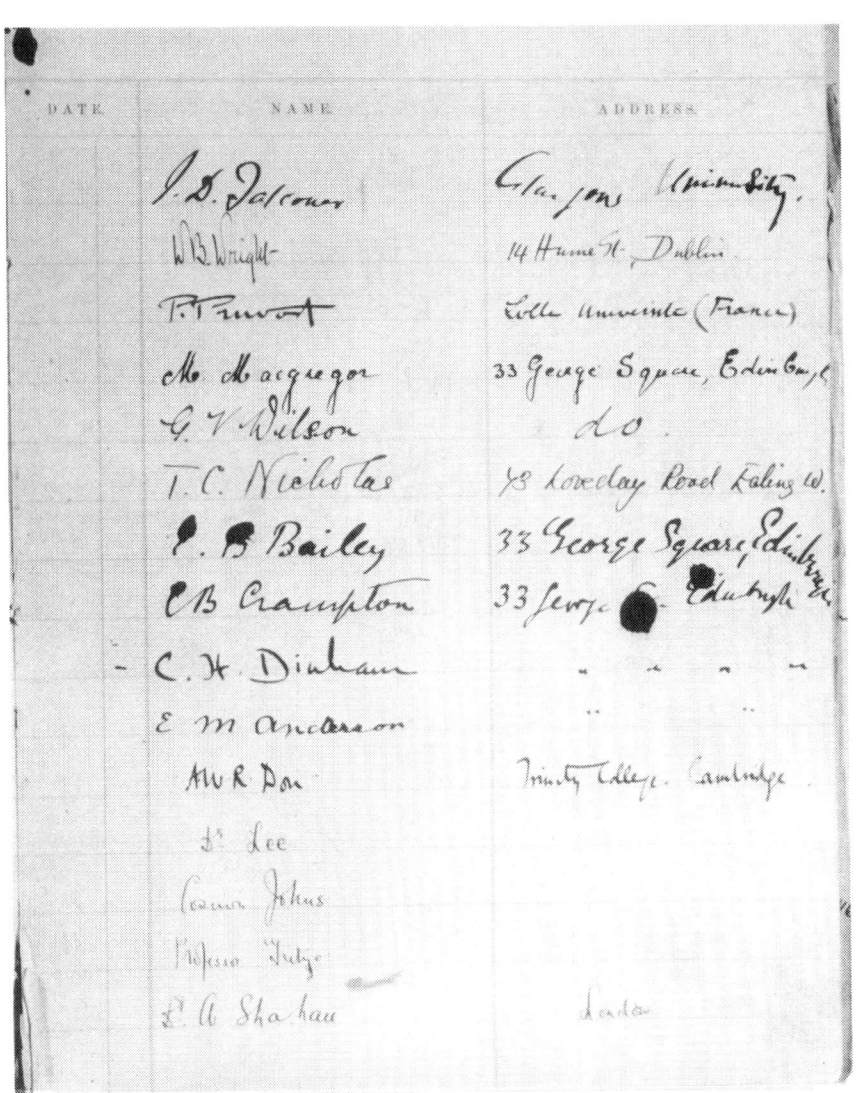

DATE	NAME	ADDRESS

Plate 6. Page in the Inchnadamph Hotel Visitor's Book for 12th-18th September 1912.

Plate 7. The participants in the 1912 BA excursion at the Inchnadamph Hotel. (Photograph by Professor S H Reynolds.)

1.	Dr J Horne	16.	Dr E E A Tietze
2.	Dr H H Reusch	17.	Dr E Jérémine
3.	C H Dirham	18.	Prof G E Haug
4.	A Gilligan	19.	Dr B N Peach
5.	C Johns	20.	Dr A Heim
6.	Prof J D Falconer	21.	Mme Barrois
7.	J E Richey	22.	Prof M Lugeon
8.	C B Crampton	23.	Prof C E Barrois
9.	E B Bailey	24.	M Macgregor
10.	J C Nicholas	25.	P E Pruvost
11.	Prof W S Boulton	26.	W B Wright
12.	Prof C K Leith	27.	G V Wilson
13.	Dr A Strahan	28.	Dr G W Lee
14.	Dr T J Jehu	29.	W F P McLintock
15.	E M Anderson	30.	A W R Don

Key to plate 7.

THUMBNAIL SKETCHES OF THOSE ATTENDING
THE 1912 EXCURSION (See plates 5, 6 and 7)

PEACH AND HORNE

The names Benjamin Neeve Peach (1842-1926) and John Horne (1848-1928) are inseparably linked as a result of their outstanding work together on the problems of the Southern Uplands and the Northwest Highlands of Scotland. A large number of maps, joint publications in learned journals of the day, and the two huge memoirs, *"The Silurian Rocks of Britain - Vol.1"* (1899) and *"The Geological Structure of the Northwest Highlands of Scotland"*, (1907) were the results of their collaboration.

Peach's father, Charles, had impressed Murchison with his knowledge of fossils, and had drawn his attention to fossils in the Durness Limestone. Young Ben Peach joined the Geological Survey from the Royal School of Mines in 1862, probably on the recommendation of Murchison, and when Horne joined five years later, he had the good fortune to receive his early training from Peach. This was the beginning of a lifelong friendship and co-operation in research, which is probably unique in the annals of science. They became known to their colleagues as Castor and Pollux, the Heavenly Twins.

Peach retired from the Survey in 1905 as District Surveyor, and initially spent much time compiling his *"Monograph on Higher Crustacea"* (1908), and later on joint publications with Horne on diverse aspects of Scottish Geology. He was a man of great energy who retained youthful boisterousness until old age. He was a talented mapper and artist, and a skilled fisherman. The story is told of how, when in the Northwest, he would work long hours, but on each third day he would take time off and go fishing.

Peach had an intuitive approach, quickly solving complex geological problems in the field. This did not always stand up to careful scrutiny, but the moderating influence of John Horne's logical thinking and careful systematic work made their partnership ideal, and proved immensely beneficial to Highland geology. Left to his own devices, Peach might have published little. He was elected FRS in 1892 and received an honorary degree of LLD from the University of Edinburgh.

John Horne became Assistant Director (Scotland) in 1901, and retired from that position in 1911. He continued to collaborate with his colleagues, contributing to many later publications. He had hoped to write a comprehensive study of the geology of Scotland along with Peach, but undertook such an extraordinary range of activities that this work was never completed. He wrote a large number of papers on various aspects of geology, and the bibliography of his published works runs to some 140 items, 48 of which were done in co-operation with Peach. He served on the councils of many learned societies and was elected FRS. He was President of the Royal Society of Edinburgh, the Geological Societies of Edinburgh and Glasgow, and of the Royal Scottish Geographical Society. He received many honours from home and abroad, including that of honorary LLD from three Scottish Universities.

From 1886, Peach and Horne were much in demand to lead excursions in Assynt, and their names appear several times in the Inchnadamph Hotel register.

JEHU, Dr Thomas John (1871-1943): Lecturer in geology at the University of St Andrews

Jehu graduated in medicine and science at Edinburgh and then in geology at Cambridge, taking up his post at St Andrews in 1903. He became Professor of Geology at Edinburgh in 1914. In 1912 he made the important discovery of fossils of Upper Cambrian age in the Highland Border rocks. After further investigations in this area he turned his attention to the Inner and Outer Hebrides, and undertook detailed mapping of this huge and complex area during his vacations over a number of years. The result was the first full description of the geology of these parts - a truly striking contribution to geological science. He had many contacts with various scientific societies: he was elected FRSE in 1906 and was President of the Edinburgh Geological Society in 1917-18.

BARROIS, Charles Eugène (1851-1939): Professor at the University of Lille, France

Barrois obtained doctorates in zoology and geology at Lille, where he researched the connection between the Cretaceous formations of France and England. After further researches in Spain he embarked on his great work, the mapping of the complex geological formations of Brittany which occupied him for twenty-nine years. For this he produced twenty geological maps representing an area of over 25,000 km^2. He became Director of the Institute of Geology in 1902. Thereafter he revised the survey of the Franco-Belgian coal basin, and founded a mining museum and school in Lille, which embraced all aspects of the coal industry. He gained international prestige for detailed works on stratigraphy and paleontology and received many honours.

STRAHAN, Dr Aubrey (1852-1928): Assistant to the Director of the Geological Survey of Great Britain

On leaving Cambridge University in 1875, Strahan joined the Geological Survey, and rose to become the Director in 1914. He was an excellent field geologist, and worked mainly on the coalfields of South and North Wales and of England, and also in the Isle of Wight. His tact, wide interests and capacity for administration made him an ideal Director during the difficult war years when there were pressing demands for information. He instigated the new series of *"Special Reports on the Mineral Resources of Great Britain"*, the first serious attempt by the Survey to react quickly to a national need. He was President of the Geological Society of London for 1912-14, was elected FRS, and was knighted in 1919 before retiring in 1920.

LEITH, Charles Kenneth (1875-1956): Professor at the University of Wisconsin, Madison, USA.

After graduating in 1897 Leith took just six years to rise from Assistant Geologist with the US Geological Survey to become Professor of Geology. He researched the structure of the Appalachian Mountains and problems of Precambrian rocks in general, before concentrating on the economic resources of the US. He was largely responsible for convincing the government of the importance to the nation of economic geology, and was adviser to a number of Presidents. He published 200 papers and fourteen books, including *"The Precambrian Geology of North America"* (1909) and *"The Geology of the Lake Superior Region"* (1911). He was president of many learned societies, including the Geological Society of America, and received many honours. He often recounted with pleasure his visit to Assynt in 1912.

REYNOLDS, Sidney Hugh (1867-1949): Professor at the University of Bristol.

Initially Reynolds taught geology and zoology at Bristol, becoming Professor of Geology in 1910. On his retirement in 1933 he was made Professor Emeritus. After researches on the igneous rocks of Southwest Ireland, his main work shifted to Southwest England, where he gradually concentrated on formations of Carboniferous age. Reynolds was engaged in research for almost fifty years and his publications were correspondingly numerous. He maintained his interest in igneous rocks and zoology, investigating the granites of Southwest Scotland and publishing text books on zoology and Pleistocene mammals. He was a very keen photographer, and his work in providing innumerable sets of prints and lantern slides for teaching purposes is recognised as a most important contribution to the teaching of geology. It is to him that we owe the photographs of the 1912 members. He was secretary of the BA Geological Photographs Committee from 1910 to 1947, and over the years he presented some 1200 prints to their collection.

McLINTOCK, William Francis Porter, BSc (1887-1960): Curator of Geology, Royal Scottish Museum, Edinburgh.

After graduating from Edinburgh University, McLintock started in 1907 as Assistant Curator in the Museum of Practical Geology in London before moving to Edinburgh in 1911. He worked mainly on gemstones and on the zeolites of the Mull lavas, gaining his DSc for the work. In 1921 he moved back to London as Curator of the Museum of Practical Geology and was involved in the early assessment of gravitational and magnetic surveying. From 1930 much of his time was taken up planning the move of the museum to South Kensington, where it opened in 1935. He became Temporary Director of the Survey in 1936 until the appointment of E B Bailey and when Bailey retired in 1945 he became Director, retiring in turn in 1950. His term in that office was largely taken up with rehabilitating the Survey and Museum after the Second World War. He was awarded a CB in 1951.

RICHEY, James Ernest, BA (1886-1968): Geologist with the Geological Survey in Edinburgh.

Richey became a District Geologist in 1925, working in many areas of Scotland. Among the many official publications bearing his name, the Ardnamurchan Memoir and map gained him world-wide recognition. He worked mostly on the Carboniferous and Tertiary successions and on the Moine Schists. After leaving the Survey in 1946, he remained active as a consultant and lecturer at Dundee University. He was elected FRSE in 1927, FRS in 1938, awarded DSc from Trinity College Dublin in 1945, and won many honours from learned societies. He was President of the Geological Society of Glasgow 1929-32 and of Edinburgh 1946-48.

BOULTON, William Savage (1867-1954): Professor of Geology, University College, Cardiff.

After his early geological training, Boulton was Assistant Lecturer and Demonstrator to the great Professor Lapworth at Birmingham University for seven years before moving to Cardiff as Professor. He returned to Birmingham to succeed Lapworth as Professor when the latter retired in 1913. His teaching expanded the Lapworth style of emphasising field observations and mapping, which had brought a complete innovation to geological instruction in the 1880s and 90s. He published a six-volume

"Text Book of Practical Coal Mining", and stressed the national importance of geological research on coal. He also did major work on water resources. His public lectures stimulated the recognition of the importance of economic geology.

HEIM, Dr Albert (1849-1937): President of the Geological Commission of Switzerland.

Heim studied under Escher von der Linth, the father of Alpine geology. Escher died in 1872 and Heim succeeded him as Professor of Geology at Zürich Polytechnic. From 1875 he also held that post at Zürich University. Heim was a lucid and prolific author. His *"Mechanisms der Gebirgsbildung"* (1878) spread Escher's ideas on how the Alps were formed, and became the authoritative work on Alpine tectonics. His *"Geologie der Schweiz"* (1916-22) is possibly the finest national geology ever produced. He was a gifted artist, illustrating his writing and lectures with his own drawings and models. Although Heim relinquished his professorships in 1911 he remained President of the Geological Commission of Switzerland until 1926. His many honours and awards included the Freedom of Zürich, given in 1899.

GILLIGAN, Albert, DSc (1874-1939): Assistant Lecturer in Geology at the University of Leeds.

Gilligan was a pioneer in the techniques of heavy mineral analysis, and his main work was on the petrology of the Millstone Grit. He gained his DSc for his research work in 1918 and became Professor of Geology at Leeds in 1922. He served for many years on the Council of the Yorkshire Geological Society, and was its President in 1929. He was an active excursion leader, and was a very popular public lecturer on a wide variety of scientific topics.

LUGEON, Maurice (1870-1953): Professor at the University of Lausanne, Switzerland.

Lugeon worked principally in the Alps, and also in the Carpathian and Tatra mountains. He published major texts on alpine tectonics, the earliest of which updated Heim's theories, and showed for the first time the interrelationships of a vast series of recumbent folds and thrust sheets. This memoir, *"Les grandes nappes de recouvrement des Alps du Chablais et de la Suisse"* (1901), became a source-book for the new science of tectonics, and gained Lugeon an international reputation by the time he was thirty. His theories have since been applied to all parts of the world. His work as a consultant made heavy inroads into his time, but he continued making detailed surveys of the Alps, which in themselves became major contributions to tectonics. He composed and set to music *"La Chanson du Moine Thrust"* which the company sang on their last evening in Assynt.

HAUG, Gustave Emile (1861-1927): Professor at the Sorbonne, University of Paris.

In 1897 Haug moved from the University of Strasbourg to the Sorbonne as a lecturer and became full professor in 1911. His scientific activity was immense and diverse, and his lecturing superb. He organized the famous geological reference museum at the Sorbonne and this formed the background for his monumental *"Traité de geologie"* (1907-11) which rapidly became one of the indispensable reference works of geology. He wrote outstanding memoirs on palaeontology, and unravelled the formation of the Rhone Basic and of geosynclines in general. His interpretations of the structure of the Alps and other mountain ranges were a turning point in the understanding of mountain building. In 1903 he discovered the great overthrusts of the Alps, confirming the proposals made by Bertrand twenty years before. He received many honours and awards and was President of the French Geological Society in 1902.

JEREMINE, Dr Elisabeth [née Tschernaieff] (1879-1964): Lecturer in Petrology, University of St Petersburg, Russia.

After working as an assistant at the University of St Petersburg Jérémine did her doctoral thesis on the Swiss Alps in Lausanne under Lugeon. She returned to Russia where she worked on the petrology of the Kola Peninsula. In 1917 she left Russia under a false name in the guise of a governess and settled in France where Haug engaged her as a lecturer in petrology at the Sorbonne. She became a French citizen and later transferred to the museum which was gaining an international reputation as a reference centre. She continued to lecture, and her dedication to her chosen science and the museum are legendary. She co-operated with workers world wide on petrological problems and was author of some 140 papers which described rocks from all over the world. In her later years she took a particular interest in the study of meteorites. She carried on working and died in harness at the age of 85.

FALCONER, John Downie (1876-1947): Professor of Geography at the University of Glasgow.

Falconer was initially assistant to James Geikie, Professor of Geology at the University of Edinburgh, before taking the post of Principal of the Mineral Survey of Nigeria with the Colonial Office. While there in 1911, he wrote *"The Geology and Geography of Northern Nigeria"*. In that same year he was invited to take up the Chair of Geography at Glasgow, a position he held until 1916, when the British Government asked him to return to Nigeria in 1916, and when they set up the Geological Survey of Nigeria in 1981, he became its first Director. He continued in this post until retiring in 1927, working on the economic resources of the country, particularly the tinfields. He then moved to Uruguay, where he was Geologist to the Government until 1934, and published reports on various aspects of the country's geology.

WRIGHT, William Bourke, BA (1876-1939): Senior Geologist with the Irish Geological Survey in Dublin.

After training at Trinity College, Dublin, Wright joined the Geological Survey of Great Britain and Ireland and soon encountered the two main interests of his career: ice ages and Carboniferous geology. In 1906 he transferred to Scotland where he helped unravel the volcanic complexities of Mull. Then followed ten years with the Irish Survey (which was now a separate entity) and the publication of his great work *"The Quaternary Ice Age"* (1914). In 1921 Wright was invited to England as District Geologist charged with setting up an office in Manchester and revising the Lancashire coalfield. This he did with great success and his research on the Coal Measures brought him a DSc from his old university in 1928. Despite hostility from his geological contemporaries, Wright supported Wegener's theory of continental drift and was one of very few British geologists to do so.

PRUVOST, Pierre Eugène (1890-1967): Assistant Preparator at the Coal Museum of the University of Lille.

Pruvost's museum work involved assisting Professor Barrois in his studies of the North French Coalfield. His doctoral thesis (1919) on the Coal Measures impressed the Belgians - who invited him to study their coalfield - and his university, which made him Professor of Applied Geology in 1922. In 1926 he succeeded Barrois as Professor of Geology and Mineralogy at Lille. By then his research had broadened to include the geology of Brittany and of Jurassic strata. His academic career, crowned by ten years as Professor of Geology at the Sorbonne, ran alongside service as Coalfield Administrator of the Lorraine Basin, a post he held from 1948 until his death. He was also the founder and President of the International Congress of Carboniferous Geology and Stratigraphy and a President of the International Geological Congress's Subcommission of the Lexicon of Stratigraphy.

MACGREGOR, Murray, MA BSc (1884-1966): Geologist with the Geological Survey in Edinburgh.

Macgregor's early experience in the Highlands, the Midland Valley and on mineral resources led to his being put in charge of work in the Northern Highlands and in the Scottish coalfields in 1919. He became Assistant Director (Scotland) in 1925. On retiring in 1945 he took an appointment with the National Coal Board as Divisional Geologist, from which he retired in 1957. As the acknowledged expert on Scottish Carboniferous stratigraphy and mineral resources, he remained a consultant until 1964. He was elected FRSE in 1922, was President at different times of both the Glasgow and Edinburgh Geological Societies, and Vice President of the Geological Society of London. He was author or part author of some eighty papers.

WILSON, George Victor, BSc (1886-1960): Geologist with the Geological Survey in Edinburgh.

After working on the Carboniferous rocks of Ayrshire and on the Mineral Resources of England and Scotland, Wilson joined the team completing the survey of Sutherland. He became District Geologist in 1928 and supervised the mapping of Orkney and Shetland. Further surveys of North Skye were interrupted by the war years, when he again became involved in mineral resources work. He retired a few years early on health grounds.

NICHOLAS, Tresillian Charles, BA (1888-1989): Research Assistant at the University of Cambridge.

For his work on North Wales Nicholas was elected as a Fellow of Trinity College in 1912, but his researches were interrupted when he had to take teaching jobs to help keep his family. During World War I he served as a surveyor with the Royal Engineers and was awarded the MC and OBE. Returning to Cambridge after the war he lectured in the Department of Geology and gradually became deeply involved in administrative duties. He was made responsible for the running of the Sedgwick Museum, became Senior Bursar of Trinity College, Chairman of the Board of Geography and Geology and a member of the Council of the Senate and of the University Finance Board. During this time he continued with research in the Lake District. On his 100th birthday, he recalled attending the 1912 excursion to Assynt and how the party sang *"La chanson du Moine Thrust"* in the Inchnadamph Hotel.

BAILEY, Edward Battersby, BA (1881-1965): Geologist with the Geological Survey in Edinburgh.

After distinguished war service Bailey became District Geologist in west Scotland until 1929, when he resigned to become Professor of Geology at Glasgow University. He returned to the Survey in 1937 to serve as Director until his retirement in 1945. At first he worked mainly on igneous and metamorphic rocks and was concerned with fourteen 'one-inch' maps and memoirs, the Glen Coe and Mull memoirs becoming classics. He had an encyclopaedic knowledge of tectonic research, and re-assessed the structures of most mountain systems. He received many awards both at home and abroad, and was elected FRSE in 1920, FRS in 1930 and was President at different times of both the Glasgow and Edinburgh Geological Societies. He was knighted on retiring from the Survey in 1945.

CRAMPTON, Cecil Burleigh, MB CM (1871-1920): Geologist with the Geological Survey in Edinburgh.

Crampton worked mainly in the Midland Valley, Glasgow, Caithness and Ross-shire. He made the important discovery of relict sedimentary structures preserved in the Moines in the aureole of the Carn Chuinneag granite in Ross-shire. He retired early, in 1914.

DINHAM, Charles Hawker, BA (1883-1955): Geologist with the Geological Survey in Edinburgh.

Initially Dinham worked on the metamorphic rocks of Sutherland and on the Midland Valley Coalfields. During World War I he was involved with mineral resources work. He became District Geologist in charge of surveying the Fife and Kinross coalfield in 1922 and was transferred to England in 1927 to be in charge of the Midlands and Cambridge Unit. He contributed to memoirs on Carboniferous geology, but his forte was his meticulous accumulation of data on British geology, much appreciated by his younger colleagues. He carried on working after normal retiring age, even when he was no longer drawing a salary. His colleagues considered the minute and precise detail of his 6-inch field maps was not excelled even by those of Clough, a geologist renowned for his precision in mapping.

ANDERSON, Ernest Masson, MA BSc (1877-1960): Geologist with the Geological Survey in Edinburgh.

Anderson contributed extensively to our knowledge of the geology of Scotland. In a major or minor degree he contributed to seven Highland Memoirs, seven coalfield Memoirs and four Lowland District Memoirs. He became Senior Geologist in 1922 but retired because of ill health in 1928.

While his early publications constitute an impressive record, it was only after his retirement that he was able to make full use of his outstanding mathematical ability, and gained an international reputation in geophysics. He wrote extensively on the dynamics of faulting and all aspects of igneous intrusion as well as optical theory. He was elected FRSE in 1922, became a DSc of Edinburgh University in 1922, and gained many awards for his outstanding work.

DON, Archibald William Robertson, BA (1891-1916): Research Assistant at the University of Cambridge.

Don did important work at Cambridge on the Lower Old Red Sandstone fossil plant, *Parka decipiens*. Soon after the Assynt excursion he left geology to study medicine. During World War I he served as a Lieutenant in the Black Watch and died in Greece.

LEE, Dr Gabriel Warton (1880-1928): Geologist with the Geological Survey in Edinburgh.

After gaining his DSc from the University of Geneva, Lee carried out investigations into the deep-sea deposits brought back by the *Challenger* expedition. In 1907, because of his expertise in palaeontology, he was invited to join the Survey in Edinburgh. He later became Senior Geologist in charge of the Palaeontological Department. He was an authority on Cambrian and Mesozoic fossils and made valuable contributions to the Survey Memoirs of the Edinburgh, Glasgow, East Lothian, North Ayrshire, Mull, Golspie and Ardnamurchan areas. He was responsible for the memoir *"Mesozoic Rocks of Applecross, Raasay and northeast Skye"*. In addition to his work for the Survey, he described suites of fossils collected in the Arctic by various expeditions.

JOHNS, Cosmo (1866-1951): Lecturer in Mechanical Engineering in Sheffield.

Johns became Superintendent of Vickers Ironworks in Sheffield, and was an enthusiastic amateur geologist. He devoted his leisure time to the study of the Lower Carboniferous rocks in the Skipton and Malham areas of Yorkshire. He served on the Council of the Yorkshire Geological Society, and contributed to a handbook on the geology of the Sheffield area.

TIETZE, Dr Emile Ernst August (1945-1931): Director of the Geological Survey of Austria, Vienna.

After studying at the Universities of Breslau and Tübingen, Tietze joined the Austrian Geological Survey in 1870, becoming Chief Geologist in 1885,

Director in 1902, and retiring in 1918. Much of his early work was in surveying the Carpathian Mountains, then part of Austria, and he published memoirs on a wide variety of topics from this area, including that of overthrusting. During work in the Elburz Mountains of Persia, he was able to explain the stratigraphy and tectonics, recognising the gradual formation of the range by repeated folding. He investigated oil problems, and as early as 1879 advocated the organic origins of petroleum. He took great interest in the erosional processes which sculpt the present day landscape. He received many honours at home and abroad.

REUSCH, Dr Hans Henrik (1852-1922): Director of the Norwegian Geological Survey in Christiania (Oslo).

Reusch gained his doctorate from Oslo University in 1883, and five years later, on the death of the Professor of Geology, succeeded him as Director of the Survey. He held this post until he retired in 1921. Although he concentrated on putting the regional geology of Norway on a sound footing, he was in later years able to revive his earlier researches on metamorphic and geomorphological problems. He was the founder and President of both the Norwegian Geological and Geographical Societies, and launched the *"Norsk geologiske Forening"* and the *"Norsk geologiske Tidsskrift"*. He started, and for many years edited, a natural history magazine and wrote schoolbooks on natural history. Reusch came from an artistic family, and many of the illustrations in his publications are in his own hand.

NB: Reusch's name does not appear in the hotel register, but that of Strahan appears twice. His name is included in other lists of the participants, and he appears in the group photographs.

BOOK AND MAP LIST

PEACH, B N, HORNE, J, GUNN, W, CLOUGH, C J and HINXMAN, L W. 1907. The Geological Structure of the Northern Highlands of Scotland. *Memoirs of the Geological Survey of Great Britain.*

BAILEY, E B. 1935. *Alpine Essays.* Clarendon Press, Oxford.

BAILEY, E B. 1952. *The Geological Survey of Great Britain.* Thomas Murby, London.

JOHNSON, M R W and PARSONS, I. 1979. *Macgregor and Phemister's geological excursion guide to the Assynt District of Sutherland.* Edinburgh Geological Society, Edinburgh.

JOHNSTONE, G S and MYKURA, W. 1989. *British Regional Geology: The Northern Highlands of Scotland.* HMSO, London.

OLDROYD, D R. 1990. *The Highlands Controversy.* University of Chicago Press, London.

CRAIG, G Y. 1991. *The Geology of Scotland* (3rd Edit.). The Geological Society, London.

SHELLEY, D. 1992. *Assynt - Geological Motor Trail.* Sutherland Tourist Board.

---ooOoo---

GEOLOGICAL MAP	BRITISH GEOLOGICAL SURVEY - One-inch special sheet (1:63360) Assynt District, Scotland.
ORDNANCE SURVEY MAP	LOCH ASSYNT - Landranger Series Sheet 15 (1:50,000)

---ooOoo---

ACKNOWLEDGEMENTS

Many people have helped in the preparation of this booklet and we should like to offer special thanks to Isles Strahan who first drew our attention to the 1912 excursion; to Mr W Morrison, the previous owner of the Inchnadamph Hotel and to Mr & Mrs Clint, the present owners, for their co-operation. The photographs of Peach and Horne, of the Peach and Horne memorial, of the group, and of Cadell's thrusting experiment are reproduced by permission of the Director of the British Geological Survey. The staff of the British Geological Survey Archives, Library and Photographic Department are thanked for their help over many months. Thanks are due to Graeme Sandeman for producing the figures and to Caroline Saunders for typing the manuscript.